Let's explore
CLIMATE CHANGE
together

RUMBLE in the Snowy Mountains
by Benita Sen

Room to Read

Room to Read seeks to transform the lives of millions of children in low-income communities by focusing on literacy and gender equality in education. Working in collaboration with local communities, partner organizations and governments, we develop literacy skills and a habit of reading among primary school children, and support girls to complete secondary school with the relevant life skills to succeed in school and beyond. Learn more at www.roomtoread.org.

No part of this publication may be reproduced in whole or in part or stored in a retrieval system, or transmitted in any form or by any means, electronic, mechanical, photocopying, recording, or otherwise, without written permission from the publisher.

Rumble in the Snowy Mountains
Author: Benita Sen
Editors: Deepali Agarwal, Laura Atkins
Designer: Christy Hale
ISBN 979-8-4000-0043-0
Copyright © 2022 Room to Read
Printed in Canada

Room to Read
465 California Street, Suite 1000
San Francisco, CA. 94104
USA

Room to Read

Table of Contents

CHAPTER ONE	Namaste!	4
CHAPTER TWO	What Is a Habitat?	6
CHAPTER THREE	Huge Himalayan Home	8
CHAPTER FOUR	Hero of the Himalayas	10
CHAPTER FIVE	The Snow Leopard and Adaptation	12
CHAPTER SIX	A Keystone Species	14
CHAPTER SEVEN	Falling Numbers	16
CHAPTER EIGHT	Climate Change	18
CHAPTER NINE	What Can We Do?	22
	Glossary	24

1 Namaste!

Hello, friends!
Have you been to the Himalaya mountains?
The air is crisp and cold.
The wind whips around with a whoosh.
Mists swirl.

The Himalayas are the tallest mountains in the world!
Very few people can climb so high.
Fewer are lucky enough to spot the snow leopard.

What Is a Habitat?

'**Habitat**' means home. It is where a plant or animal lives. The right habitat has the conditions needed for life. These include air, food, water, and soil. Every creature in a habitat is important. Even the tiny ant!

Forests cover almost one-third of the land on Earth. They give us **oxygen**, fruits, and medicines, and help to purify the air.

FOREST HABITAT

Oceans cover around 70% of the Earth's surface. They keep the Earth at a comfortable temperature to live.

OCEAN HABITAT

DESERT HABITAT

Deserts cover more than a fifth of land on Earth. A variety of animals and plants have adapted themselves to the harsh conditions of cold and hot deserts.

6

MOUNTAIN HABITAT

The Himalayas are a mountain habitat. Temperatures can be very cold. As you go higher, the soil becomes thinner and the air has less oxygen to breathe. Plants and animals need to be tough to survive here.

③ Huge Himalayan Home

The Himalayas are the habitat for millions of plants and animals. About 53 million people live here. Any disturbance in the Himalayas will harm them all.

Are the Himalayas important? Yes! Without them, surrounding lands would not have enough water. Do you know why?

Glaciers are large areas of thick ice. They are found in the highest part of the Himalayas. When glaciers melt, the water feeds rivers that flow down the mountains. This water is very important for plants, animals, and people.

HOME TO BIRDS
The Himalayas are home to more than 900 **species** of birds. The Himalayan monal is a pheasant. To fight the winter cold, large groups of monal huddle together in winter.

HOME TO TREES AND PLANTS
Many plants grow in the Himalayas. The Himalayan blue poppy is used to make medicine. It is a rare plant.

HOME TO ANIMALS
You would recognise the red panda from its reddish-brown coat and long, ringed tail. In spite of its name, it is not related to the giant panda that lives in China.

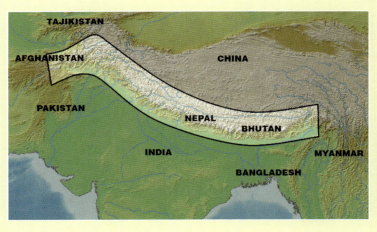

The Himalayan mountain range crosses a large part of Asia. The melting snows provide water to many countries, including India.

4 Hero of the Himalayas

Now let's meet the hero of the higher Himalayas, the snow leopard! The snow leopard gets its name from the snowy mountains that are its habitat. It usually lives among rocks and shrubs, though it can also be found in grasslands and even forests.

The snow leopard has beautifully patterned skin. It mostly moves about alone, during dawn and dusk.

The snow leopard needs to hunt a mountain goat or sheep every week or 10 days. However, it eats whatever it can hunt, from blue sheep and deer to rodents and even birds. There are about 4,000 to 8,000 snow leopards left in the world. Of those, between 250 and 1000 make their home in India.

Scientists have only recently started to track snow leopards. That is why it is hard for us know exactly how many are left, and where they live. The snow leopard also keeps moving from place to place, and is hard to spot!

5 The Snow Leopard and Adaptation

Animals have **adaptations**, or special skills, that help them survive in their habitat. These adaptations could be changes to the body or in their behavior.

Snow leopards have thick fur coats to stay warm. Their coloring makes them difficult to see against the rocks.

The snow leopard's powerful build allows it to go up and down the steep slopes easily. Its hind legs are so strong that it can jump a distance equal to six times its own length.

The snow leopard has smaller ears to keep in more body heat. Larger ears would also flap in the icy wind!

The longest, thickest tail among big cats allows snow leopards to balance on the rocks. It also works like a muffler.

Broad paws keep the snow leopard from sinking into soft snow. Their paws are lined with fur to keep in heat and for better grip on steep slopes. The snow leopard can move on rocky ground and deep snow.

A Keystone Species

Every habitat has its own food supply. In the Himalayas, almost no other animal hunts the snow leopard. It is one of the largest animals there.

snow leopard

ibex

goat

partridge

If the animal at the top disappears, then the number of animals it feeds on will grow.

The snow leopard is so important that it is called a 'keystone species'. It holds its habitat together like a keystone supports an arch. If a keystone species is in danger, its habitat and all of the life there will also be in danger.

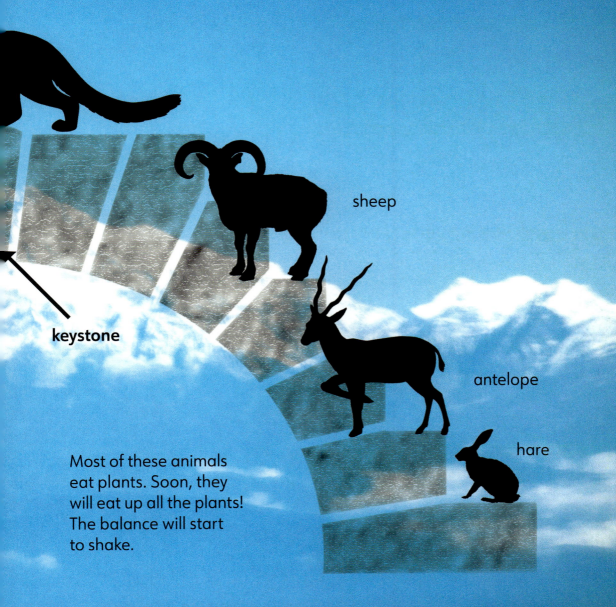

keystone

sheep

antelope

hare

Most of these animals eat plants. Soon, they will eat up all the plants! The balance will start to shake.

7 Falling Numbers

Is the snow leopard okay? No. In the recent years the population appears to be falling in many places. Scientists fear that it may go **extinct**.

The snow leopard is perfectly adapted to the cold conditions of Himalayas, but its habitat is shrinking. The region is growing warmer and has less snow. This causes people to move their fields of crops higher in the mountains. The snow leopard's habitat shrinks even further.

Human activities like mining, and building roads and dams disrupts its habitat. This forces the snow leopard down the mountains where villagers hunt it out of fear. The Indian government launched Project Snow Leopard to **conserve** the animal and its habitat.

Why are these things happening?
Climate change is a very big reason.

Scientists leave camera traps to try to count snow leopards and see where they are living.

8 Climate Change

Human actions have caused a rapid rise in global temperatures. People have burned **fossil fuels** such as coal, oil, and gas to make energy for over two hundred years. When they burn, these fuels release a gas called **carbon dioxide**. Carbon dioxide traps extra heat on the planet.

Cutting down forests also releases carbon dioxide. Every tree absorbs carbon dioxide

Can you find the climate change effects that are mentioned in this book?

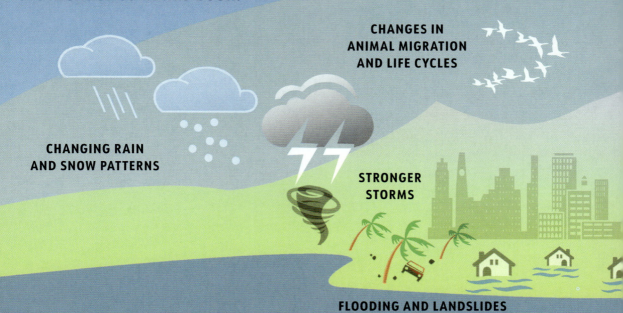

CHANGING RAIN AND SNOW PATTERNS

CHANGES IN ANIMAL MIGRATION AND LIFE CYCLES

STRONGER STORMS

FLOODING AND LANDSLIDES

RISING SEA LEVEL

as it grows. Trees help to keep the level of carbon dioxide in proper balance.

This rise in global temperature is changing **weather** patterns all over the world. This is called climate change. Many plants, animals and humans are struggling because of these new weather patterns.

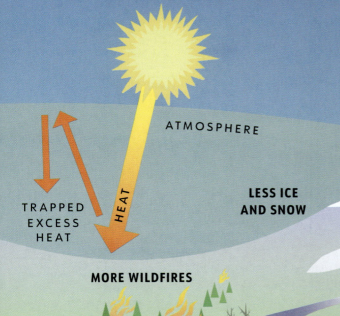

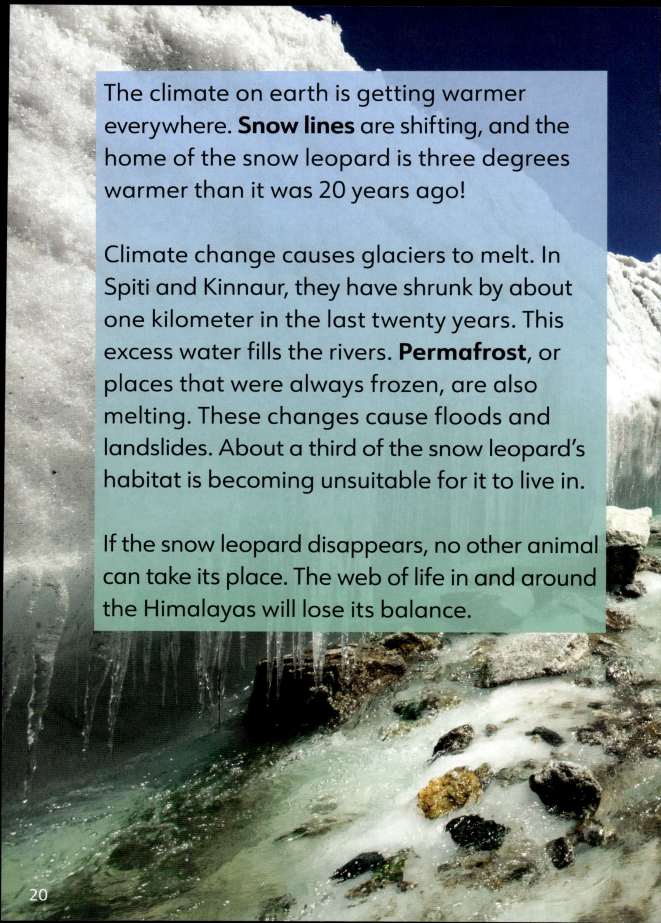

The climate on earth is getting warmer everywhere. **Snow lines** are shifting, and the home of the snow leopard is three degrees warmer than it was 20 years ago!

Climate change causes glaciers to melt. In Spiti and Kinnaur, they have shrunk by about one kilometer in the last twenty years. This excess water fills the rivers. **Permafrost**, or places that were always frozen, are also melting. These changes cause floods and landslides. About a third of the snow leopard's habitat is becoming unsuitable for it to live in.

If the snow leopard disappears, no other animal can take its place. The web of life in and around the Himalayas will lose its balance.

9 What Can We Do?

You may be wondering what to do about climate change and how to help the snow leopard. Here are a few ideas.

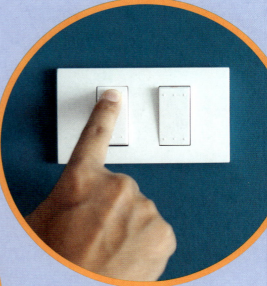

Switch off when you leave a room. Electricity is one reason for climate change, so the less we use, the better for the Earth.

Read about the snow leopard and the problems it faces due to climate change. Share what you learn with others.

Each of us can make a difference. We also need to work together with our families, communities and government leaders to help fight climate change.

Try to walk or bike rather than go by car. Most cars run on fossil fuels like petrol and diesel that add to climate change.

Use reusable materials like cloth bags. Producing plastic bags adds to climate change. Also, plastic does not degrade and pollutes our land and sea.

Glossary

Adaptation: The process in which a living thing changes to exist in a particular environment

Carbon dioxide: Often referred to by its formula CO_2, this gas traps the heat from the sun in the Earth's atmosphere

Climate change: The long-term changes in the Earth's weather patterns

Conserve: To protect or keep safe from being changed or destroyed

Extinct: A type of plant or animal that no longer exists

Fossil fuels: Fuels such as coal or oil that come from old life forms that decompose over millions of year

Habitat: Home for a certain plant or animal

Oxygen: A gas that is found in air and water, and is needed for people, plants and animals to live

Snow line: The level on mountains above which snow never melts completely

Species: A group of plants or animals that are all the same kind

Weather: Natural conditions of a place at one time, including temperature, sunlight, rain, clouds, and snow